FACT FINDERS
KNOW THE DIFFERENCE!

REPTILE OR AMPHIBIAN?

EZRA E. KNOPP

PowerKiDS press™

Published in 2026 by The Rosen Publishing Group, Inc.
2544 Clinton Street, Buffalo, NY 14224

First Edition

Editor: Greg Roza
Book Design: Michael Flynn

Photo Credits: Cover (chameleon) Alex Popov/Shutterstock.com; cover (frog) Kurit afshen/Shutterstock.com; p. 4 Marc Freebrey/Shutterstock.com; p. 5 (salamander) Milan Zygmunt/Shutterstock.com; p. 5 (toad) David Havel/Shutterstock.com; p. 5 (turtle) Bildagentur Zoonar GmbH/Shutterstock.com; p. 6 Alessandro Bruno Varanini/Shutterstock.com; p. 7 reptiles4all/Shutterstock.com; p. 8 Nitissak Pisil/Shutterstock.com; p. 9 Ondrej Prosicky/Shutterstock.com; p. 10 EugenioHMPhotography/Shutterstock.com; p. 11 HWall/Shutterstock.com; p. 12 Heiko Kiera/Shutterstock.com; p. 13 RJCMiddleton/Shutterstock.com; p. 14 W. de Vries/Shutterstock.com; p. 15 Irina Furashova/Shutterstock.com; p. 16 Pk_camera/Shutterstock.com; p. 17 abs galeri/Shutterstock.com; p. 18 Patrick Hatt/Shutterstock.com; p. 19 feathercollector/Shutterstock.com; p. 21 (giant tortoise) Dr. Juergen Bochynek/Shutterstock.com; p. 21 (axolotl) Lapis2380/Shutterstock.com.

Cataloging-in-Publication Data

Names: Knopp, Ezra E.
Title: Reptile or amphibian? / Ezra E. Knopp.
Description: Buffalo, New York : PowerKids Press, 2026. | Series: Fact finders: know the difference! | Includes glossary and index.
Identifiers: ISBN 9781499452815 (pbk.) | ISBN 9781499452822 (library bound) | ISBN 9781499452839 (ebook)
Subjects: LCSH: Reptiles–Juvenile literature. | Amphibians–Juvenile literature.
Classification: LCC QL644.7 K56 2026 | DDC 597.9–dc23

Manufactured in the United States of America

CONTENTS

WHICH IS WHICH?

Reptiles and amphibians are two of the main groups of animals on Earth, and they have many similarities, or things in common. Both are omnivores, which means they eat both plants and animals. They are both cold-blooded, which means their body temperature is affected by their surroundings. So many similarities can make it hard to tell which is which!

Scientists have studied both types of creatures for many years. Through observation and investigation, they have discovered how these creatures are very different from each other.

SNAKE

DO YOU KNOW WHICH OF THESE ANIMALS ARE REPTILES AND WHICH ARE AMPHIBIANS?

TURTLE

TOAD

SALAMANDER

IN THE WILD

Reptiles can be found in many **environments** around the world. Sea turtles spend most of their lives swimming in Earth's oceans, coming onto land only to lay their eggs. The thorny devil lizard thrives in the deserts of Australia by collecting water with its thorny skin!

Amphibians can be found in many of the same environments as reptiles. However, they need to live in areas that have a dependable source of water. This might be in a creek, a **swamp**, or even under the ground.

SCALES OR MUCUS?

When many people think of reptiles, they think of scales. Scales are dry, hard plates that help keep a reptile's body safe from harm. Some reptiles, such as snakes, have thin, flat scales. Other reptiles, such at alligators, have thicker scales called scutes.

Do amphibians have scales? Far from it! Amphibian skin is smooth, soft, and covered in **mucus**. Mucus helps amphibians stay moist, or wet. It may also help amphibians escape from predators that can't hold on to them.

TAKE A BREATH

Reptiles, just like mammals such as people, have lungs. Lungs are **organs** that take in a gas called oxygen from the air and release a gas called carbon dioxide. This process is called pulmonary respiration. "Pulmonary" means having to do with the lungs. "Respiration" means breathing.

CAIMAN

SOME SALAMANDERS, SUCH AS THE DUSKY SALAMANDER SHOWN HERE, DON'T HAVE LUNGS OR GILLS! THEY ONLY BREATHE THROUGH THEIR SKIN.

All baby amphibians are born in water. They use **gills** to get oxygen. Many adult amphibians have simple lungs. However, most adults are able to breathe through their skin! This is another reason why they need to stay moist—to keep breathing!

BABY ON BOARD!

Most reptiles lay eggs with thick, leathery shells. The shells keep the growing baby reptiles safe from predators and the environment. The number of eggs a reptile lays depends on the species, or kind. Reticulated pythons, shown here, can lay up to 80 eggs at one time.

Some reptiles stay with the eggs to help keep them safe, but others leave them as soon as they are done laying them. When the reptiles break out of their shells, they look like tiny **versions** of the adult reptiles.

Amphibians lay eggs too. However, amphibian eggs, as well as the amphibian **life cycle**, are very different from those of reptiles. Amphibians only lay their eggs in water. The eggs are surrounded by mucus to keep them safe.

TADPOLES

When baby amphibians break out of their eggs, they don't look like adults. Baby frogs, for example, are called tadpoles. They are tiny blobs with tails. In time, tadpoles grow and change into adult frogs. This change is called metamorphosis.

TONGUE TIED

You can tell amphibians and reptiles apart by observing their tongues. Amphibians generally have a fat tongue covered with a sticky mucus. This helps them to pick up prey. Some, such as and toads, can shoot their tongue out to grab prey!

Many reptiles have a bifurcated, or forked, tongue. You've probably seen a snake or lizard flick out its tongue. They do this to pick up the smell of nearby prey! A forked tongue helps them tell where the smell is coming from.

WHO'S BEEN AROUND LONGER?

You may have heard that reptiles such as crocodiles and alligators "walked with the dinosaurs." It's true! Their earliest **ancestor** appeared about 200 million years ago when dinosaurs did indeed walk the Earth.

THIS IS THE SKULL OF AN ALLIGATOR ANCESTOR NAMED DEINOSUCHUS THAT LIVED ABOUT 80 MILLION YEARS AGO.

Knowing this, which do you think have been on Earth longer—reptiles or amphibians? Amphibians first appeared about 370 million years ago! Their earliest ancestor was a species of fish similar to modern lungfish, which have lunglike organs and can breathe air.

KEEP INVESTIGATING!

This is just the start to what you can discover about reptiles and amphibians. Did you know that axolotls don't go through metamorphosis like other amphibians? They are born with gills and lungs and keep them their whole lives. Did you know that tortoises can live much longer than people? The oldest reptile in the world is a giant tortoise named Johnathan!

Knowing what you've learned through careful observation and investigation, can you tell which of these animals is a reptile and which is an amphibian? It's not that hard!

THE STUDY OF REPTILES AND AMPHIBIANS IS CALLED HERPETOLOGY. THIS COMES FROM THE GREEK WORD *HERPETÓN*, WHICH MEANS "CREEPING ANIMAL."

GLOSSARY

ancestor: Someone in your family who lived long before you.

chemical: Matter that can be mixed with other matter to cause changes.

environment: The conditions that surround a living thing and affect the way it lives.

gills: A body part that fish and some amphibians have that allows them to breathe in water.

life cycle: A series of changes that take place as a living thing grows.

mucus: A slippery, sticky substance produced made in the body that moistens and protects parts of the body.

organ: A body part that does a specific job.

poison: A harmful kind of matter that can kill a living thing.

swamp: Land that's always wet and often partly covered with water.

version: A form of something that is often different from the ones that came before it.

FOR MORE INFORMATION

BOOKS

Baker, Kalena. *What Do You Think About Reptiles? Decide For Yourself*. Charleston, SC: Teaching Made Practical, 2025.

Jackson, Tom. *The Magnificent Book of Reptiles and Amphibians.* San Rafael, CA: Weldon Owen, 2025.

WEBSITES

Reptiles

kids.nationalgeographic.com/animals/reptiles

Amphibians

kids.nationalgeographic.com/animals/amphibians

National Geographic Kids has much more to say about reptiles and amphibians!

INDEX